THE
CORPUS
CLOCK

Published by:

Fromanteel Ltd

Arragon Mooar

Church Rd Santon

Isle of Man IM4 1HB

British Isles

British Library Cataloguing in Publication Data

A catalogue record for this book is available from the British Library

ISBN 978-0-9548339-4-7

Text by Christopher de Hamel, paintings of Matthew Parker and Marlowe and
the leaf from the Apocalypse are reproduced by the kind permission of the
Master and Fellows of Corpus Christi College, Cambridge.

The portrait of John C Taylor is reproduced courtesy of Louise Riley-Smith.

Illustrations on pages 13 & 41 by Alan Meeks   www.visitechdesign.com

Art direction and design by Tyra Till +44 (0)1538 300401
info@tyratill.com

Printed and bound in Great Britain by BAS Printers   www.basprint.co.uk

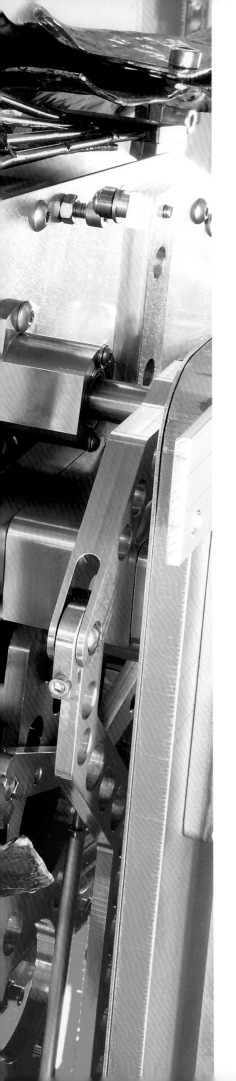

by Christopher de Hamel

with photography by John C Taylor

THE
CORPUS
CLOCK

THE CORPUS CLOCK

The Corpus Clock, a unique and strange device for the measurement of time, is both hypnotically beautiful and deeply disturbing. It was invented, designed and given to Corpus Christi College, Cambridge, by Dr John C Taylor, principal benefactor of the Taylor Library, the new undergraduate library of the College, opened to its readers in 2008. The Taylor Library now occupies a large part of the Victorian gothic building at the junction of Trumpington Street, King's Parade and Bene't Street, Cambridge, originally constructed in 1867 as the London and County Bank, and until recently in use as a branch of the National Westminster Bank. The building has always belonged to Corpus and, when the bank's lease expired in April 2005, the College reclaimed it and began construction of the Taylor Library. Since the main entrance into the Library is from within the College site, leading into it from what is now called Library Court, the former nineteenth-century street entrance became redundant and was closed off. The old stone doorway of the former bank now provides the frame for the Corpus Clock.

The clock is a remarkable mixture of very modern design and an ancient setting; of precision engineering and engaging whimsy; of utterly traditional clockwork (quite literally) and unexpected electronic innovation; of vast size and extreme delicacy of movement; of unceasing life and imminent death; and it tells the time with absolute exactness and breathtaking unpredictability.

Time and death have been themes of Corpus Christi College since the beginning. The College was founded in 1352 as a result of the Black Death, the plague which killed nearly a third of the population in England in the mid-fourteenth century. Two of the town guilds, or associations of local tradesmen, those of Corpus Christi and of the Virgin Mary, joined together to purchase this site, then on the edge of the city, to create a new college to help attract business back into Cambridge, severely depopulated by the plague. It was thus a municipal foundation, unique among the medieval colleges of Oxford and Cambridge, most of which were established by individual benefactors, often aristocratic or royal, or were ecclesiastical houses, originally for the training of monks. Corpus was prominent in law and medicine, as well as theology. The site included the Anglo-Saxon church of Saint Bene't, the oldest building in Cambridge, with its graveyard, and for over 200 years the church served as the College's chapel. The first scholars prayed there for the living and the dead. For part of its history the College was popularly known as Bene't College, but the original name of Corpus Christi and the Blessed Virgin is more accurate and is in use today. Even the words 'Corpus Christi' have a certain theme of mortality, for they reflect the late medieval veneration of the Holy Sacrament and the dead Body of Christ, as it was in the hours between the Crucifixion and Easter, three days later. Death and the passage of time are ancient themes here.

The arms of the College of Corpus Christi and the Blessed Virgin Mary appear in bas relief on the clock. They were granted to the College in 1570, at the request of Matthew Parker. The devices are the pelican in her piety with the three lilies of the Annunciation. In medieval iconography the pelican was a symbol of the Body and Blood of Christ, 'Corpus Christi' in Latin, because the bird was reputed to shed its own blood to save its offspring. The lilies were emblems of the purity of the Virgin Mary at the moment of the Annunciation.

The medieval churchyard of Saint Bene't is a reminder of the transitory nature of life.

The original part of Corpus Christi College is not visible from the street and is worth seeing. Enter by the main gates in Trumpington Street into the early nineteenth-century New Court, and go left through either of the passageways on the north side, emerging on the far side into what is called Old Court, a virtually intact quadrangle from the second half of the fourteenth century. It is small and agreeably self-contained. On the facing wall is a sundial, with the Latin motto which translates as 'The world and its desires pass away' (we will come back to this text later), and there is a plaque recording the residence as students here of two major Elizabethan playwrights, Christopher Marlowe in 1581–87 and John Fletcher in 1591–94. If one stands back, the tops of two buildings can be seen emerging over the speckled roof tiles of Old Court. Ahead is the early eleventh-century tower of Saint Bene't's Church, with its tall round-arched Anglo-Saxon windows. To the right, however, are the six gables of the former Cavendish Laboratory, where the atom was first split (1917) and where DNA was discovered (1953). Early medievel Christianity, therefore, stands within a few feet of the beginning of the nuclear age.

There was a library in Corpus Christi College from the start. It was probably in the rooms on the upper floor on the south corner of the east range of Old Court. There are some manuscripts still in the possession of the College which have been here since the Middle Ages, including a small illustrated Apocalypse, on the imminent end of time, bequeathed to Corpus by Thomas Markaunt in 1439. The greatest library benefaction, however, was

Sundials controlled the pace of medieval life as clocks were rare, inaccurate and expensive.

The Old Court and Cavendish Laboratory.

4

The Markaunt Apocalypse
(Parker Library, MS 394).

The only reputed contemporary portrait of **Christopher Marlowe** is in the possession of the College.

that of Matthew Parker (1504–1575). Parker himself lived here in Old Court, first as an undergraduate (1521–25) and a Fellow (1527) and again as Master of Corpus from 1544 until 1553, when he was deprived of his post in the reign of the Catholic Queen Mary. In 1559 Queen Elizabeth brought him back from exile and made him Archbishop of Canterbury, with instructions to make the English Reformation absolute and irrevocable. Looking for historical justification and precedent for an independent English Church, Parker gathered into his own possession the oldest manuscripts he could find in the country, mainly from the suppressed monasteries and the remnants of libraries in the medieval cathedrals, now converted to Anglicanism. Parker eventually entrusted about 400 manuscript volumes and several thousand early printed books into the care of Corpus Christi College, with stringent conditions of preservation and scholarly access. The collection includes many monuments of supreme importance, such as the sixth-century Gospels of Saint Augustine, the original of the Anglo-Saxon Chronicle, and the early twelfth-century Bury Bible. In the 1820s Parker's library, still intact, was moved into an elegant upstairs room in what is now called New Court,

The south-east corner of **Old Court**, with the rooms where Matthew Parker lived when Master.

designed by the architect William Wilkins. It is one of the finest small collections of medieval books in the world.

In the late 1990s, Corpus Christi College resolved to secure the long-term future and safety of this collection. For much of the twentieth century the College's undergraduate library, comprising the daily working resources of the students, was housed immediately beneath the Parker Library, and both parts of the collections had by then reached a desperate crisis of space. The imminent availability of the former bank building, on the outer corner of the College site, suggested a solution, which was to create a new student library there and to move out the modern books from the ground floor of New Court. This would in turn vacate space for a proper vault for the medieval manuscripts and the formation of a new rare books reading room and research centre.

This statue of **Matthew Parker** stands outside the Chapel.

This portrait by Louise Riley-Smith shows **Dr John C Taylor** sporting one of his four Queen's Awards.

Attention therefore turned to the student library. At this point Dr John C Taylor entered the story. He was born in Buxton in Derbyshire, attended school on the Isle of Man, and came up to Corpus Christi College as an undergraduate in 1956, finishing a B.A. (later M.A.) in Natural Sciences in 1959. By occupation, John Taylor is an inventor. He now lives once again on the Isle of Man. Many of the hundreds of patents which he holds are connected with domestic appliances, thermostats, and electrical equipment. His single most famous invention is the cordless kettle, patented and used now throughout the world. Sales have been phenomenal. John Taylor holds four Queen's Awards, three for Export (1995, 1998 and 2002) and one for Innovation, given in 2000 for his 360° cordless kettle connector. He became a Fellow of the Institute of Patentees and Inventors in 1965 and he received an honorary doctorate in Engineering in 2000. In 2002 he was elected to an Honorary Fellowship of Corpus Christi College. As the plans for the new library began to unfold, Dr Taylor offered to sponsor a large amount of the cost, and it was named the Taylor Library in his honour.

Throughout much of his life, John Taylor has been immersed in the study and collecting of early clocks. His other personal interests include mountaineering and flying (he first went solo in 1953 and he still pilots his own planes). He has organised major exhibitions on seventeenth- and eighteenth-century horology, with a special interest in the works of Christiaan Huygens (1629–1695), inventor of the pendulum clock, and John Harrison (1693–1776), inventor of the marine chronometer. John Taylor established Fromanteel Ltd., a horological development company, named after the Fromanteel family of clock makers of seventeenth-century London. Clocks offer a rare marriage between science and art, and the complexity of mechanism and utter simplicity of concept clearly link the inventions of John Taylor himself today with those of his horological predecessors of the past. Once the working mechanics of a clock are in place, the instrument itself can become a medium for art, sometimes of great elaboration, and when the clock begins to tick, with a sound like a little heartbeat of its own, it seems to be a living thing. Engineering, artistry and natural science all appear to come together in clocks.

John Taylor in Reykjavik, piloting a Citationjet.

John Taylor has often observed that there have been no significant changes in the mechanical clock since the eighteenth century. For some years, therefore, he has been working on a new form of clock, a traditional timepiece, driven by a spring, as in the past, and paced by a rocking escapement, as always, but which measures and shows time in an altogether original and innovative way. It is the first full-sized example of this new kind of clock, which looks out appropriately over the city of Cambridge from the former entrance of the building now called the

The creation of the **Taylor Library** has led to the formation of a new and graceful Library Court, in a space previously occupied by bicycle sheds and a laundry block.

Taylor Library, and from the College founded next to
St Bene't's Church by the guilds of the town.

   Let us consider the outside of the Corpus Clock first,
before examining the mechanics of its inner workings.
First of all, its massive face is plated in pure gold, polished
to a mirror finish. It is possible to see in it fragmented
reflections of the street and world behind the viewer. It
is made to resemble radiating ripples, as if a stone had
dropped into the middle of a pond of liquid metal. It was
created by a series of explosions in a vacuum, pounding

The face is massive and remarkably robust, and is formed from a single thin sheet of high quality stainless steel, about 1500 mm (or five feet) in diameter.

the hard metal into shape. The ripples allude to the greatest explosion of all, the Big Bang, the central impact which formed the universe and the beginning of time, sending out pulsating waves which are still in motion. Time is undulating outwards in waves and troughs of gold.

Above the universe and dominating the clock is an extraordinary monster. Unlike the clock face, the creature is mostly black, with a hint of blood and an apparent dusting in powdered gold; and it is clearly organic, moving up and down, and back and forth. It is spiky and ugly (and disconcertingly modern), a jarring contrast to the silent gold planisphere below. Its lacy metallic wings quiver eerily, and its tongue lolls in its ghastly mouth. Perhaps the nearest literary parallel is one of the locusts from the bottomless pit in the Apocalypse (Revelation 9:3).

This diagram shows the wooden front pallet of Harrison's grasshopper escapement engaging the escape wheel teeth exactly as in the photograph opposite.

This diagram shows the wooden rear pallet performing the same function as the rear leg of the Chronophage in the photograph opposite.

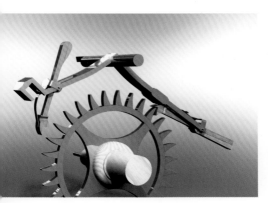

Dr Taylor uses the word 'Chronophage' for the beast on the clock, a rare but authentic word meaning 'time-eater', for that is what it seems to be doing, devouring each minute as it passes. It resembles no earthly creature but it evolves out of a grasshopper, a term used by John Harrison himself to describe his invention of an escapement which was a strictly functional innovation. Harrison's original 'grasshopper escapement' rocked back and forth, driven by the power of the escape wheel below. As it lifted itself up and fell back between the teeth of the wheel, once every second, it actually restrained and measured out the speed of rotation.

Harrison's innovation of the grasshopper eliminated the problem of sliding friction, since it had two points of contact with the teeth on the escape wheel – in effect, a front and back leg – each in turn moved the wheel fractionally backwards and thereby released the other, first one and then the other, caused by the 'grasshopper' rocking backwards and forwards. There were no sliding parts where the feet and the teeth interacted, merely a touch and release, and so it was less likely to wear out. More importantly, no oil was needed to lubricate the interaction of components. This was crucial, because oil thickens and thins out according to fluctuations in temperature. Even small variations in viscosity would affect the consistency of the measurement of time. That is precisely what the creature on the Corpus Clock does and, although it has a symbolic value too, the primary reason for its presence is utterly mechanical. It functions as the Harrison grasshopper does. It engages its feet onto the teeth of the escape wheel below by means of pallets, which end in little horizontal bars between the monster's toes.

An ornamental pointer descends from the bottom of the bob, with a small adjustable weight for regulation. This weight was a common feature in seventeenth-century clocks, and can be screwed up and down to make fractional adjustments to the rate of the swing. The tip of the pointer hovers above a graticule, shaped like a long highly polished metal dish. This is plated in gold and in rhodium, the silvery white metal more precious than gold. It is sculpted with four oval hollows on either side of the ornamental arms of Corpus Christi College in its centre.

The ridges where the hollows touch each other mark the graded tenths of a second. The tip of the pointer above indicates each tenth as it passes, and the globe is reflected on its transit over the graticule.

The Corpus Clock has no hands, or digital numbers. Time is shown by concentric orbits of what appear to be flashing blue lights, darting or progressing at different rates around the circumference of the clock face. In fact, the lights are neither moving nor flashing on and off. They are there all the time, whether we see them or not. Day and night are like that in reality: the sun does not go out; it is simply obscured.

Within the core of the clock are three concentric rings of tiny electronic LED bulbs, arranged in 48 rows of 22 bulbs to indicate the hours, 60 rows of 16 bulbs to indicate the minutes, and 60 rows of 12 bulbs to indicate the seconds. It sounds extraordinarily wasteful to run so

The printed circuit boards mounting the **LED bulbs**.

The ten indices on the **graticule** take account of the variable speed of the pendulum, fastest in the middle of the swing, slowing to a stop at the end of each swing.

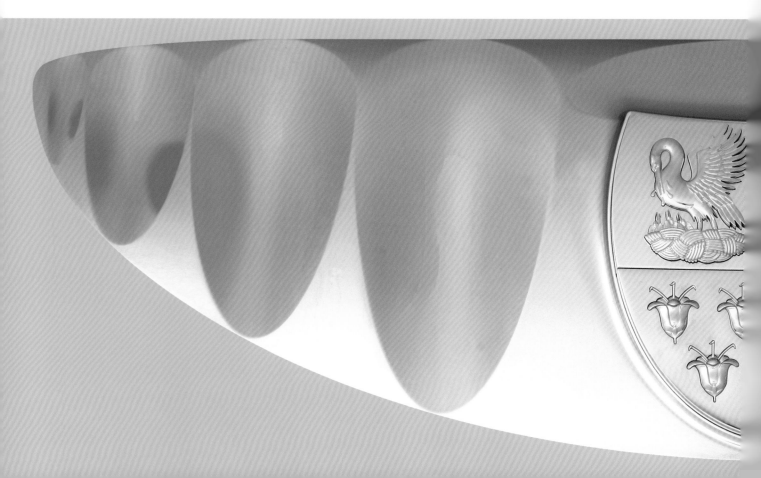

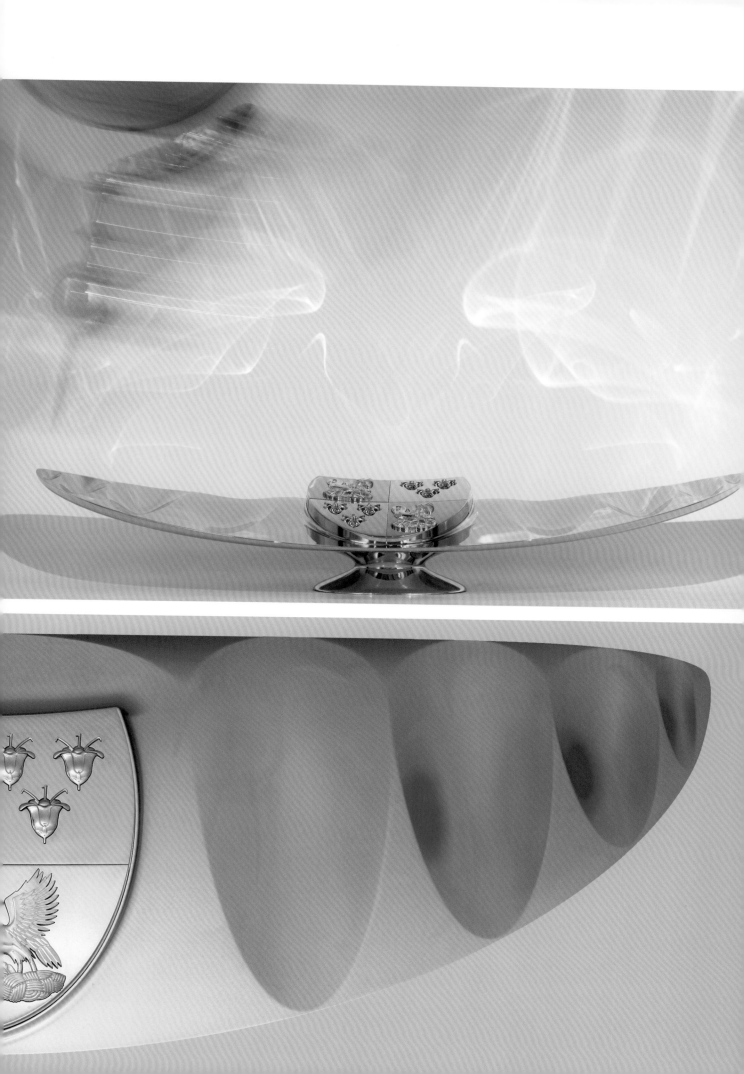

many bulbs simultaneously, but in fact the total wattage of all of the 2,736 LEDs is less than three standard 60-watt light bulbs. Even with the motor continuously winding the remontoire spring, the entire clock runs on less than 400 watts altogether.

The vast clockwork escape wheel, which moves around the outer periphery of the clock, is attached or connected by cogs to three concealed stainless steel rings which revolve independently of each other in the narrow space between the lights and the inside of the clock face. These concentric rings are punctured by elaborate sequences of radial slits. As they rotate over the lights, the slits line up with apertures in the clock face and, at any moment, one light in each ring appears to have flashed on. If this represents the cycle of seconds, each light appears to come on for a sixtieth of a second. If the light represents the cycle of minutes or hours, the mechanism is calculated so that the actual minute or hour remains illuminated.

The teeth visible around the perimeter of the Corpus Clock are attached to the edges of a huge ring or escape wheel which rotates around the edges of the clock face, measured out by the Chronophage above. The teeth are made to resemble something organic, more like the arched spines along the back of a dinosaur than the conventional teeth of a cog wheel. The teeth move around the edge of the universe in little jerks, in accordance with the movement of the monster's feet, making a sound like the primitive beat of an animal's heart. The poetic symbolism suggests that the Big Bang, an event of pure physics, sends out wave after wave of echoes which evolve, in time (vast but quantifiable time), into life on earth at its extremities, even if savage and prehistoric.

The escape wheel is actually powered by a spring, a proper old-fashioned clock spring, in the upper right-hand corner within the clock. This pushes the wheel round continuously in a clockwise direction. It is what is called a 'remontoire' spring, one which is kept constantly wound up, in this case by electricity. As the escape wheel turns, paced by the interaction with the monster's feet, its rate of rotation is modulated by the swinging of the pendulum. Once again, this is completely traditional, on the principle of the pendulum patented by Huygens in 1657. The arm of the pendulum is inside the clock, but its vast spherical terminal, or bob, protrudes below, swinging backwards and forwards, one second to the left, one to the right. The word 'bob' hardly conveys the sense of size. It is massive and weighty. Like the clock face above, it is plated with a mirrored surface of gold. It is bathed in a blue nimb from behind, like the sunlit earth in its atmosphere seen from space, in perpetual motion.

The remontoire spring, coiled inside the upper gear wheel, provides the force to drive the escape wheel forward.

The blue aura is caused by light escaping sideways from light-emitting diodes, or 'LEDs', inside the clock. If this is the terrestrial globe, as it seems to be, note that one can always see oneself reflected in the very centre of it, whether one stands to the side or directly in front of it. Mankind is at the centre of our world. The teeth appear to be sawing into the earth, whether we like it or not.

The apertures in the clock face are fitted with clear optical lenses, elliptical in shape and about a quarter of an inch wide. These transmit the blue-coloured light out from the hidden LED bulbs. The lenses protrude outwards beyond the flat surface of the clock, like crocodile eyes, enabling the illuminations to be visible even when viewed at a slight angle. It would be stretching the analogy with the universe too far to see the lenses as stars or planets, but there is certainly something astronomically appropriate about measuring time by watching orbits of celestial lights.

The development model shows that each **lens** acts as a light pipe to collimate the light from the line of LEDs mounted on the printed circuit board.

The outside – the visible clock face – has three concentric rings of apertures for lenses. The outer and middle rings, those for seconds and minutes, each have 60 evenly-spaced lenses, and the inner ring, showing the hours and quarter hours, has 48. These are the same numbers as the LED bulbs inside, as we have seen. The corresponding steel rings revolving in front of the bulbs, however, have either 61 slits or 49 slits. Obviously they

The clock with the **face removed**.

The three rings of lenses in position viewed from inside the clock.

The New Court clock.

could not have 60 or 48 slits, for then every one of the lights would appear to be on, or every one off, at any moment. If the escape wheel had 59 and 47 slits, however, the lights would appear to be moving round the clock face backwards.

Note that the basic arrangement of the clock face is completely conventional. It marks the traditional twelve numbered hours, from midnight or noon, not the trendy 24-hour clock of military commands, railway timetables or cheap digital watches. To tell the time, observe the position of lights on the perimeter of the clock face, as if you were looking at the hands of a normal clock. The top represents 12 o'clock, or zero seconds or minutes past the hour. The furthest right is 3 o'clock, or fifteen seconds past the minute or quarter past the hour, and so on. With a little practice the time is no harder to tell than on any standard clock without written numbers. The nineteenth-century clock in New Court in Corpus has no actual numbers inscribed on it, and most people never notice.

The Corpus Clock shows conventional time as follows. The outer ring represents the seconds. The pendulum completes a full swing in the space of one second. In that period the blue lights seem to rush rapidly right around the outer ring. On the completion of a full cycle the lights apparently pause for a moment of time. The pause happens because the pendulum has reached the limit of its swing, and it too actually stops for a fraction of a second before it begins to swing the other way. Where that pause takes place on the clock face shows the specific second in each minute. Each new cycle pushes the pause forward by another second. It takes a moment or two to adjust one's eyes to this flashing progression, but the time advances absolutely, relentlessly, and strangely satisfyingly. The sense of time racing away is very evident.

The minutes are similar. They too move round unrelentingly but much more slowly, taking one hour to complete a full orbit of 60 minutes. The lens lit up at any moment represents the number of minutes past the hour. At the completion of every minute, however, the lights rush round the whole circle and push the time forward by one space into the next position. They stop still for 59 seconds, and then the same thing happens again. They rush round the middle ring and come to a stop at one number further on in the circuit.

There is an unexpected but intriguing point. Any circuit of the clock face here is not 60 spaces, but 61, to move the time forwards to the next position. Thus there appear to be 61 seconds in each minute, and 61 minutes in each hour. Each blue flash represents one sixty-first of a circuit. In some ways, that may be a more accurate measurement, for time actually occupies spaces between

The clock in motion.

digits, and there are 60 spaces between the numbers one to 61, and only 59 between the numbers one to 60. This is logical but no other clock in the world shows it. This is the first moment when we begin to glimpse that Dr Taylor is taking liberties with time.

The hours are shown on the innermost ring. It has 48 lenses. Every fourth lens projects from a raised mound. The more recessed lights show a quarter past the hour, half past, and a quarter to the next hour. The lights on raised mounds are the hours. Quarter to four, for example, is more apparent to the human eye if the light has moved onwards from the position for three to somewhere approaching the position for four o'clock. On the hour, once an hour (usually, anyway), the clock strikes. This happens with a great rush of lights, as bright and fast as a cascade of fireworks. For each stroke of the strike, the lights of the hour ring appear to rush backwards around a full cycle and then forwards again for another cycle. It takes exactly one second to perform a backward circuit and another second for the forward circuit. The clock actually strikes on the backward movement. Thus striking one o'clock takes a duration of two seconds, two o'clock four seconds, and so on, up to 24 seconds, almost half a minute, to strike the hour of twelve. This does not cause time to stop, for the clock then automatically catches up. The clock is programmed to be absolutely accurate every fifth minute.

So far, at least superficially, the Corpus Clock tells absolute time, even if we are left pondering how many seconds or minutes there really are in a minute or an hour. However, the games have only just begun. Time, as any metaphysician will tell us, means different things

The time is three ten precisely.

in different circumstances. We use expressions like time standing still, or taking forever; we speak of losing time, or time running away. Dr Taylor enjoys quoting Einstein's observation about relative time that fifteen minutes sitting with a pretty girl on a park bench passes in a moment, but one minute seated on a hot stove would feel like eternity.

At another time (itself an interesting phrase), we might say that a clock stands still, especially if we watch it. This is all reflected in the Corpus Clock. It does the unexpected. Just occasionally (not very often), it will seem to stop altogether, or to have moved rapidly forward so imperceptibly that one can hardly see what has happened to the intervening time. The pendulum suddenly seems to freeze, although others might not see it like that. Sometimes we could almost swear that time went backwards. These illusions appear to happen at random, and perhaps they actually do: the point is that, despite all the precision of scientific measurement, time is unpredictable. This is the first clock that has seriously attempted to take account of that.

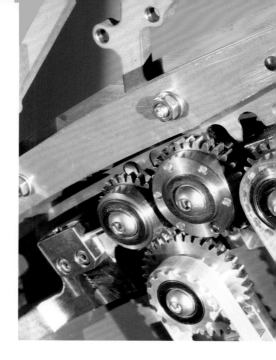

The **hour and minute linkages** from the seconds disc are activated when the six teeth drive the lantern pinion one rotation during the 59th second of every minute.

Each little tooth on this **pinion** is formed from a miniature ball race 5 mm in diameter, lubricated and sealed for life.

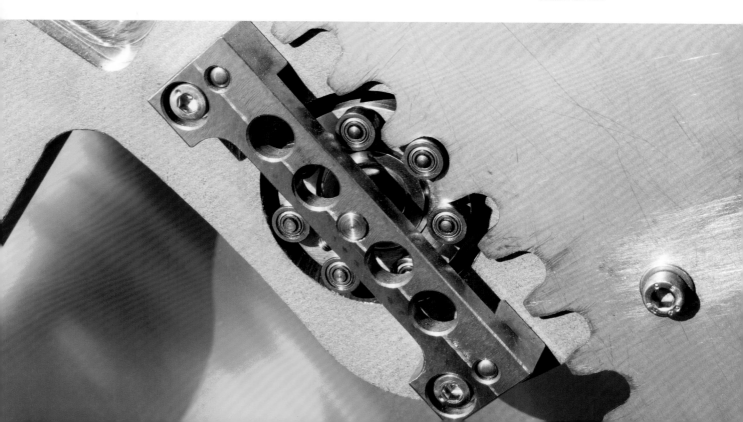

The monster, or Chronophage, on the top of the clock is the old grasshopper of Harrison, transmogrified into a very different and altogether more sinister role. It is very contemporary and its style could belong to no other period but our own. It appears as a living creature; true cosmography is inanimate and without good or evil, but the beast here has a mind and teeth. It is apparently walking forwards, and it feeds by eating time. It destroys all animate creation by taking away the time allotted to it. It looks very horrible and is meant to. Its movements

are controlled by gears and cogs inside the clock. Every 60 seconds it slowly opens its mouth and suddenly snaps it shut, devouring each minute. Its fearsome jaws begin to open about 30 seconds into the minute, reaching their full extent about quarter of a minute later. It waits in silence and, between the 59th and the 60th second, it bites; and that minute is never here again. Every fifteen minutes the sting on the creature's tail clicks up, and as the quarter hour approaches, the tail begins to quiver and move, and then it slowly sinks back. Every hour, on the hour, it snaps its mouth and shudders, shooting its sting the number of times of the hour itself. The creature's unearthly eyes blink at random, its sideways lids closing inwards.

The passing of time is not simply important because it is relentless. To human beings, time brings death. We are

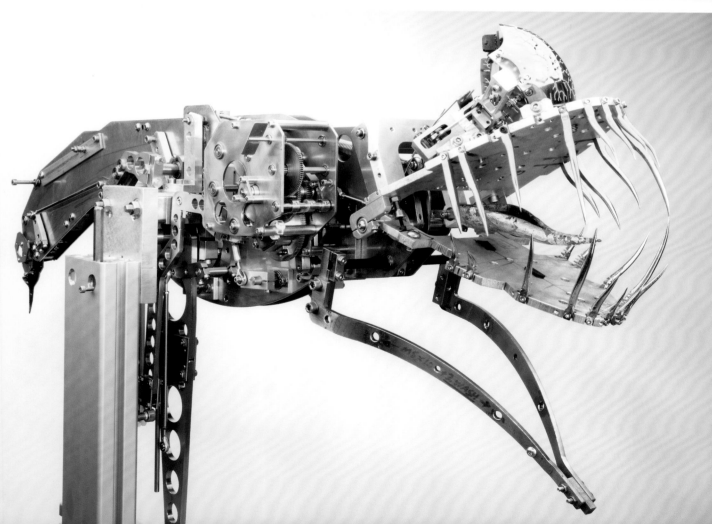

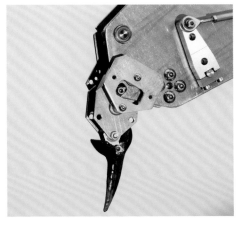

The top photographs show the **tail segments**. On the reverse side can be seen the mechanism that operates the sting.

the only living creatures aware that we will die one day. Every irrecoverable moment of time edges each one of us one measurable step closer to the grave. When the Corpus Clock strikes the hour it indicates the number not with a joyful bell, like many normal clocks, but with a

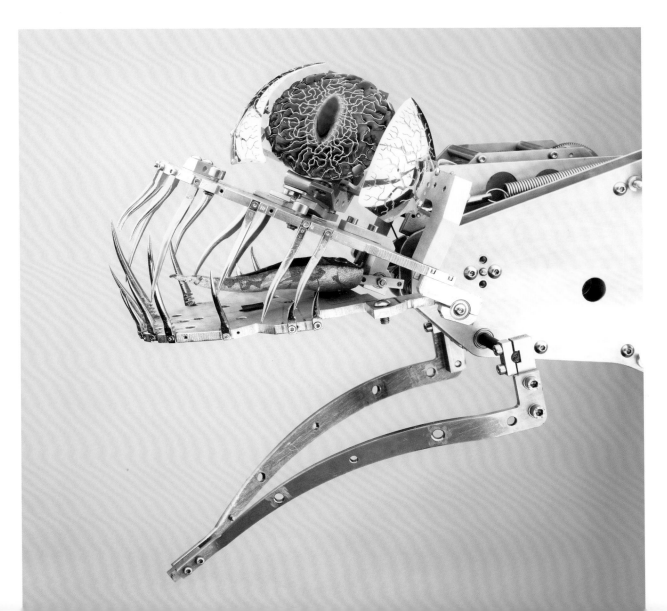

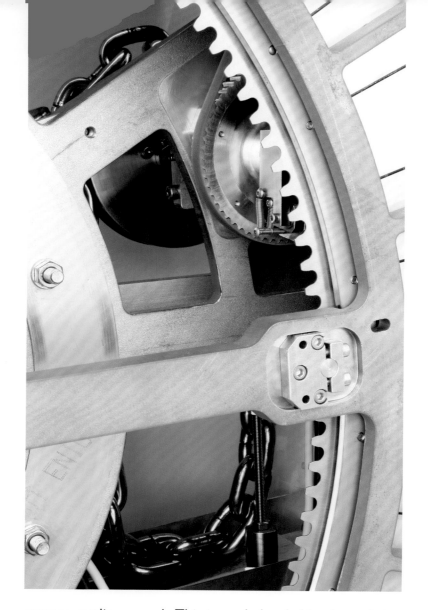

The **chain** that makes the sound and the **hammer** that strikes the wooden box concealed below.

strange rattling sound. This is made by shaking iron chains over a wooden coffin and a hammer beating the wooden lid, concealed inside the back of the clock, with the sound broadcast outwards through tiny electric speakers. It is the most eerie and disconcerting sound imaginable, and it is absolutely real. We will die.

Corpus Christi College, as explained above, was originally founded in an atmosphere of imminent mortality. This is now (understandably) an extremely unfashionable subject in our modern world, but it was second nature to people in the late Middle Ages, when plague, warfare, childbirth, medical ignorance, poor diet and lack of sanitation could cut off life at any moment. Many members of the medieval College would have lived

The **blinking eyelids** are driven by a complex clockwork mechanism.

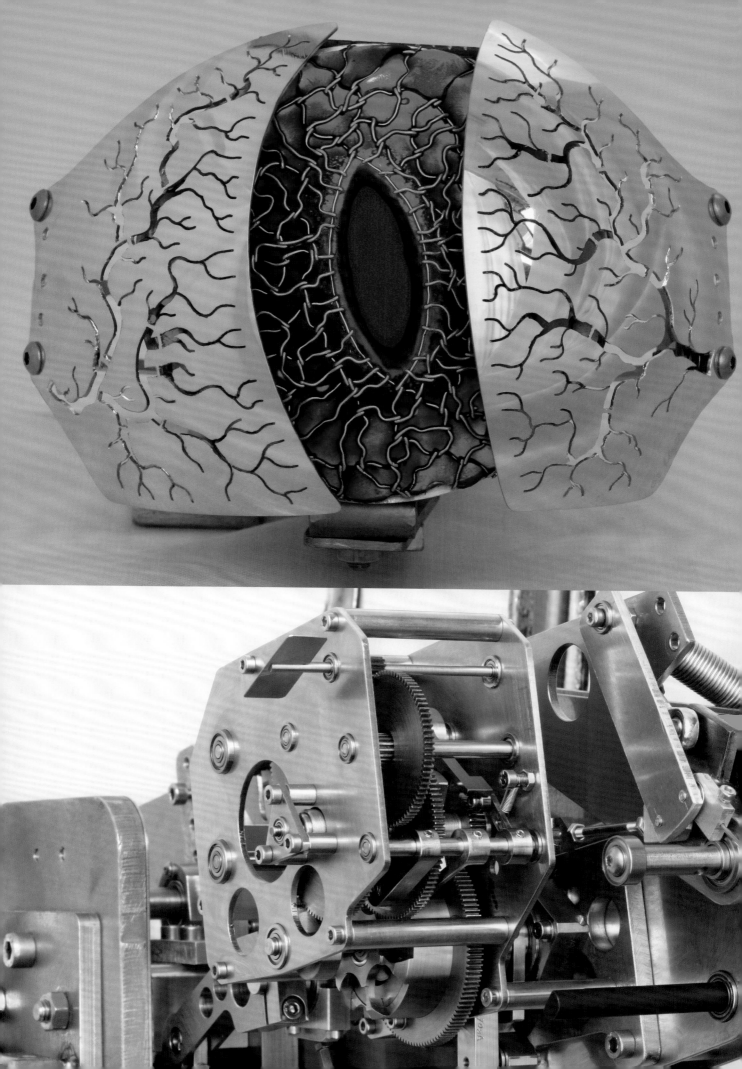

here until they died and were buried on the premises, and the chaplains would have prayed daily for the collegiate communities of the living and the dead. John Mere, who was buried in the churchyard of Saint Bene't's in 1558, bequeathed money to the College for an annual sermon, which is still preached, on a selection of suggested themes, including the imminence and inevitability of death.

In 1573 Matthew Parker had his own picture painted, probably by Richard Lyne, and it was engraved for publication by Remigius Hogenberg, one of the first engraved portraits made in England. It shows the Archbishop and former Master of Corpus at the age of 70, seated in a chair holding a Bible in his wrinkled hands, staring outwards, apparently deep in thought. On the windowsill beside him is an hourglass, with the sand running through. It is the most obvious *memento mori*, an allegory of time running out. Around the edge of the portrait is the quotation from 1 John 2:17, "Mundus transit et concupiscentia eius", 'The world and its desires pass away'. This is a kind of second motto for Parker. It is inscribed on the sundial in Old Court, and it has been cut by Lida Kindersley along the stone of two steps below the Corpus Clock, a theme which appropriately links the early College with the expression of mortality implicit in the clock.

Time, at least, is central to life in a college of a university. We speak of having the time of one's life; of one's time at College; old times; of wasting time, perhaps, or saving time; in our time; the first time for many experiences; good times; high time; term time; lecture time; one's own time; timetables; making up for lost time; no time; exam time. Most undergraduates are here for three years, and postgraduates about the same, or less. Time seems to stretch ahead unbelievably far from the lonely perspective of one's first day in an austere College room, but the creature eats time and it all passes far too rapidly, it really does, and it never comes back. The transience of those three brief years at university haunts most former students for the rest of their lives.

The greatest and most powerful literary expression of the unstoppable march of time is *Doctor Faustus*, written by Christopher Marlowe, member of Corpus Christi College. Perhaps he wrote it while in Corpus, but this is not known, for it was not performed until after his death six years later. Faustus, the ingenious university scholar, signs a pact with Lucifer for 24 years, which at the start of the adventure seems unimaginably long. It passes, and his time runs out. By scene 13, the clock is striking eleven. He declares, "Ah, Faustus, now hast thou but one bare hour to live", and he desperately commands the spheres of heaven to stand still and to let this hour become a year, a month, a week, even a single day, but in vain. The minutes go by and time refuses to stop: "The stars move still, time runs, the clock will strike." He will do anything, absolutely anything, to halt the progression of time. "Ah,

The medieval abbreviation MŪDUS for MUNDUS has been preserved in the inscription cut on the steps below the clock.

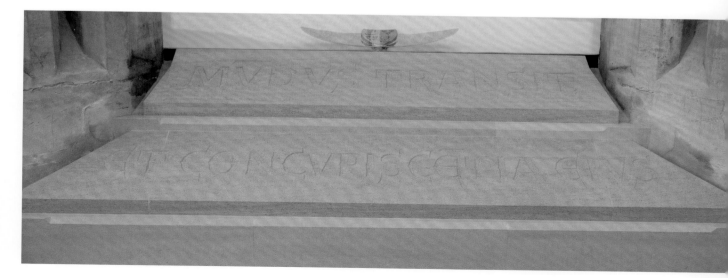

half the hour is past: 'twill all be past anon". Five minutes
remain, one minute, a second. "O, it strikes, it strikes,
now body turn to air"; there is thunder and lightning, and
the curtain falls as Faustus is dragged to Hell.

The Corpus Clock was invented and designed by
John Taylor, and was constructed principally in the
Huxley Bertram workshops under Dr Taylor's constant
supervision. The basic (and hitherto unique) concept
of marking the moments of time by concentric rings
of differently calibrated gradations derives ultimately
from the principle of a Vernier sliderule, devised by the

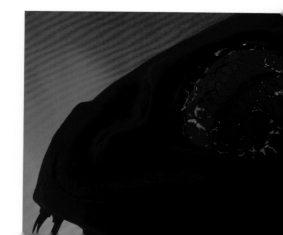

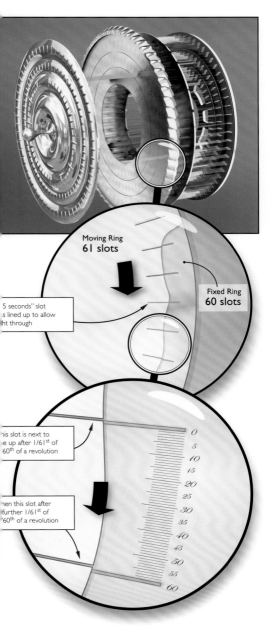

Moving Ring
61 slots

Fixed Ring
60 slots

5 seconds" slot
s lined up to allow
ht through

his slot is next to
e up after 1/61st of
60th of a revolution

en this slot after
further 1/61st of
60th of a revolution

0
5
10
15
20
25
30
35
40
45
50
55
60

The moving ring rotates 1/60th of a revolution, representing one second in time.

During this time each of the sixty slots in the fixed ring will line up with one of the sixty-one slots in the moving ring, allowing light to shine through. The slots line up, one after the other in a clockwise direction until the sixty-first slot of the moving ring (also the first slot) lines up with a slot that is one slot further on from where it started – indicating one second has passed.

mathematician Pierre Vernier (1580–1637), which uses components marked with differently spaced scales sliding up and down in parallel with each other, to define a distance by more than one calibration at once. Vernier scales are used in sextants and surveying instruments. John Taylor has adapted Vernier scales into concentric rings for the Corpus Clock. Since time is measured by more than one scale (all arbitrary), seconds, minutes, hours, and the mind's interpretation of all these and their sometimes inconsistencies with the actual cycles of the sun, it seemed an interesting way of redefining what in most mechanical clocks is primitively indicated by pointing hands, often no more sophisticated than a policeman at a traffic junction. Countless drawings and scale models were produced. Gradually it evolved into a project which would be both an innovative scientific device, with other implications, and a work of art which frankly takes science and teases it, as Einstein did, and disconcerts preconceptions about the most elusive aspect of our universe. The creation of the Corpus Clock has resulted in six new inventions and patents.

The clock and its smaller experimental prototype were both made at Huxley Bertram Engineering Ltd., high precision industrial engineering designers and manufacturers in Cottenham, near Cambridge. The specialist engineer for the project, Stewart Huxley, is in fact a great-grandson of the Victorian scientist and biologist Thomas Huxley. Others involved included Gary Moore, industrial designer from Origin Create, and Tyra Till, design consultant and project manager. The creature at the top was created from beaten copper by the sculptor Matt Sanderson. The enameller was Joan

MacKarell. The graticule, or measuring dish, was devised by Alan Meeks, of Visitech Design. The photography for this publication, and many of its captions, are by John Taylor himself.

There were many trials and challenges to be overcome in the transference from initial design to actual construction. The principal frame of the clock was cut and shaped from a single 15 mm sheet of stainless steel using jets of extremely high pressure water, a method which allowed it to be fashioned into shape, leaving strengthening ribs, reducing the overall weight considerably. The Corpus Clock is intended to be as durable and long-lasting as anything by Huygens or Harrison, for example, now three centuries old and still working. The final clock face was therefore made of stainless steel, but only a millimetre thick. Gilded stainless steel is incredibly durable. It was moulded into the required shape in an extraordinary way. Grinding it would be impossible; beating it would scar it with hammer marks. It was actually shaped by a series of five controlled explosions, carried out at a secret and specialised military site in the Netherlands (the only place in Europe where such things are possible), enclosed in a tank, with water in front of the sheet and a vacuum behind it, to diffuse the impact and to avoid possible disfigurations from the escape of air.

The pendulum is inscribed cryptically, "Joh. Sartor Monan. Inv. MMVIII" within calligraphic flourishing, resembling the statements of makers' names engraved on seventeenth-century clocks. "Joh." is *Johannes*, John, of course. "Sartor" is the usual medieval Latin word for a tailor; it survives in the adjective 'sartorial'. "Monanensis"

The problems in creating a **perfect spherical bob** and performing the engraving in traditional style tested the most sophisticated 5 axis CAM miller, and it was the last major component to be manufactured.

means from the Isle of Man (*Mona*, a word first recorded by Julius Caesar). "Inv." is *invenit*, a verb with multiple meanings, most of them appropriate: 'discovered', 'devised', 'invented', 'drew', 'contrived', 'made', or 'brought to fruition'. It announces forever, to those who gather in front of this remarkable and unpredictable invention, 'John Taylor, of the Isle of Man, made it, in 2008'.

CLOCK CRAFT

LIMBS

Lightweight smithing of copper limbs with articulated joints and enamel.

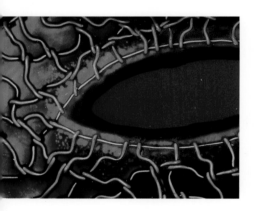

EYE

Pupil and iris cut and fused from coloured glass, slumped in a kiln over a hand-tied net of 18ct gold wire.

SPUR GEARS

Stainless steel gears from the 'drive train' between the second, minute and hour discs.

ENGRAVING

Five-axis engraving on a spherical pendulum bob.

EYELIDS

Hot-forged stainless steel. Polished and hand-pierced following an organic pattern. Gold plate.

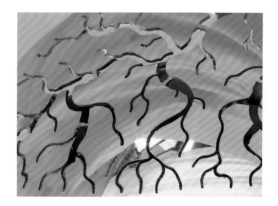

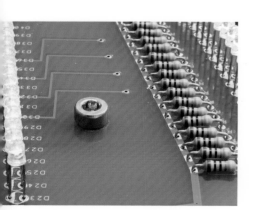

LIGHT ARRAYS

Printed circuit boards with narrow-beam blue LEDs.

FLY

Air friction 'fly' damping the eyelid motion.

TEETH

Twenty teeth hand-forged and polished from stainless steel before gold plating.

ESCAPE WHEEL TEETH

Stereo-lithographically modelled from computer drawings. These were silicone moulded to create wax patterns for lost-wax casting into aluminium, and then machined, polished and gold plated.

ENAMELLING

Multi-layered enamel finish over raised copper armour.

PLAQUE

Wire-eroded stainless steel coat-of-arms. Male and female components cut, plated with rhodium and gold before assembly.

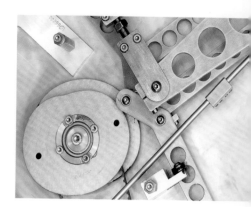

CAMS

Cam wheels controlling the palette lift.

STING

Articulated cast stainless steel 'sting' components.

LENSES

Cast acrylic lenses channelling the LED light through the face of the clock to the viewer.

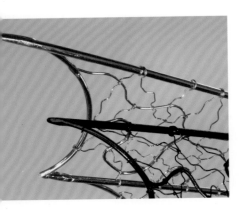

GEARS

'Lantern' gears engaging the laser-profiled slitted discs.

WINGS

Hand-tied and TIG-welded stainless steel graded wires. Gold plate and black passivate.

CENTRE DROP

CNC lathe-turned stainless steel. Gold plate.

TONGUE

Forged copper with gold leaf inclusions under crimson enamel.